NOSTALGIAWORKS

COMPANION READING

AF879886

COPYRIGHT AND PERMISSIONS NOTICE

For partnership inquiries, bulk orders, institutional sales, or permissions request, contact NostalgiaWorks at info@nostalgiaworks.com or visit nostalglaworks.com

NANA'S BOOKS, a Connecticut Limited Liability Company,
DBA NostalgiaWorks

nostalgiaworks.com

Blessed is the Nation whose God
is the Lord, they are the people
He chose for his inheritance.

Psalm 33:12

Good Housekeeping

JULY 1931

25 CENTS

Eleanor Hallowell Abbott – Fanny Heaslip Lea

Dixie Willson – Jay Gelzer – Elizabeth Frazer

Edison Marshall – Emma-Lindsay Squier – Montague Glass

May your mercy, Lord, be upon us,
as we put our hope in you.

Psalm 33

DEFEND
AMERICAN
FREEDOM
IT'S EVERYBODY'S JOB
McClelland Barclay

Therefore, take up the full armor of God, so that you will be able to withstand in the evil day, and having done everything, to stand firm.

Ephesians 6:13

Howard Chandler Christy

These things I have spoken to you,
so that in Me, you may have peace
in the world.

You have tribulation, but take courage;

I have overcome the world.

John 16:33

USA
SALUTE TO VICTORY
VICTOR TCHETCHET

Be strong and take heart,
all who hope in the Lord.

Psalm 31:24

Memorial Day

Tears are prayers, too.
They travel to God when we can't speak.

Psalm 56:8

DON'T WAIT
for the Draft
VOLUNTEER
SSOCIATED MOTION PICTURE ADVERTISERS, INC. Poster No.3
GUENTHER

When you've done everything you can do, that's when God will step in and do what you can't do.

2 Corinthians 12:10

The
WAR
CRY
CHRISTMAS
1942
NEW TESTAMENT
SALLMAN 42
© THE SALVATION ARMY, INC. 1942

He that dwelleth in the secret place of the most High shall abide under the shadow of the Almighty.

Psalm 91:1

THE AMERICAN

JUNE

LEGION

1940

MAGAZINE

"I PLEDGE ALLEGIANCE TO THE FLAG OF T OF
AMERICA, AND TO THE REPUBLIC FOR W NE

I keep my eyes always on the Lord. With him at my right hand, I will not be shaken.

Psalm 16:8

JOIN!
The Greatest Mother

God is our refuge and our strength,
a very present help in trouble.

Psalms 46:1

December 20, 1941

Soon enough I will come to know:
I am better off just one day in this timeless moment than an eternity anywhere else.
It is on this very threshold of return that I have come home.

Psalm 84:6

Thank you, Veterans!

May He remember all your sacrifices
and accept your burnt offerings.
May He give you the desire of your heart
and make all of your plans succeed.
May we shout for joy over your victory
and lift up our banners in the name of God.

Psalm 20, 3-5

TAKE IT FROM ME, BROTHER–
WE'VE STILL GOT A BIG JOB TO DO!
HOWARD SCOTT

If one member suffers, we all suffer together; if one member is honored, we all rejoice together.

1 Corinthians 12:26

For Home and Country
Alfred Everitt Orr — 18
VICTORY LIBERTY LOAN

To everything there is a season.

Ecclesiastes 3:1

YOUR BLOOD
CAN SAVE HIM
1384
WHITMAN

Be merciful to me, Lord;
for I am in distress.
My eyes grow weak with sorrow,
my soul and body with grief...

"But trust in you, Lord; I say,
'You are my God.'
My times are in Your hands."

Psalm 31:9, 14-15

"Liberty And Freedom
Shall Not Perish"
A. Lincoln
OLORED MEN
The First Americans
Who Planted
Our Flag
on the
Firing Line
TRUE SONS OF FREEDOM

And you will know the truth,
and the truth will set you free.

John 8:32

Raised 'em myself
in my
U.S. School Garden
ISSUED BY THE U. S. SCHOOL GARDEN ARMY BUREAU OF EDUCATION, DEPARTMENT OF INTERIOR,
WASHINGTON, D. C.
AMERICAN LITHOGRAPHIC CO., N.Y.

Make thankfulness your sacrifice to God, and keep the vow you made to the most High.

Then call on me when you are in trouble, and I will rescue you, and you will give me glory.

Psalm 50:14-15

GOD
BLESS
AMERICA

Have I not commanded you?
Be strong and courageous.
Do not be afraid,
Do not be discouraged,
for the love of your God will be with
you wherever you go.

Joshua 1:9

COULDN'T HAVE DONE IT WITHOUT YOU!
U.S.S. SOUTH DAKOTA
HOWARD SCOTT
Full Speed Ahead.....
WE'VE STILL GOT A BIG JOB TO DO!
OFFICIAL NAVY POSTER, INDUSTRIAL INCENTIVE DIVISION, 2—80M

Blessed is the man who remains steadfast
under trial, for when he has stood the test,
he will receive the crown of life, which
God has promised to those who love Him.

James 1:12

SHALL THIS CONTINUE?
JOIN
THE
NAVY
NAVY LEAGUE
PITTSBURGH
ANCHOR BANK
BUILDING
Morgan
LITHO. CO.
3939
CLEVELAND, U.S.A.

Help me, O Lord, my God:
O save me according to Thy mercy.

Psalm 109:26

THE CALL
TO DUTY
JOIN
THE
ARMY
FOR HOME
AND COUNTRY
ADAPTED
FROM THE
SCULPTURE BY
EDOARDO
CAMMILLI
PUBLISHED BY RECRUITING COMMITTEE OF THE
MAYOR'S COMMITTEE ON NATIONAL DEFENCE

Therefore, my beloved brothers,
be steadfast, immovable, always
abounding in the work of the Lord,
knowing that in the Lord, your labor
is not in vain.

Corinthians 15:58

SACRIFICE
THE PRIVILEGE OF FREE MEN

Offer the sacrifices of the righteous and trust in the Lord.

Psalm 4:5

To Have and to Hold!
WAR BONDS

Greater love hath no man than this, that a man lay down his life for his friends.

John 15:13

For THINE
is the KINGDOM
and the POWER
and the GLORY
FOREVER

But now I will heal and mend them.
I will make them whole and bless them with an abundance of peace and security.

Jeremiah 33:6

BOOK CHAT

LET'S FIND...

-A boy in striped socks
-A mounted soldier
-A pair of suspenders
-White gloves
-A holly wreath
-A head of cabbage
-A safety pin

FUN FACTS!

-Memorial Day was first celebrated in 1868.

-The holiday marks the unofficial start of summer.

-Memorial Day was once known as Decoration Day.

-Memorial Day honors service members who have passed.

-A moment of remembrance takes place at 3 pm on Memorial Day.

-The Lincoln Memorial was dedicated on Memorial Day in 1922.

-On Memorial Day, the flag is flown at half staff from dawn until noon.

AUTHOR SHARE:

When I was a classroom teacher, every November, my students made cards for our local veterans.
It was a wonderful experience all around. Little did I know that one day I would make books to celebrate the contributions of veterans. It is a labor of love.
Thank you for your service!

Made in the USA
Las Vegas, NV
10 November 2026

57851950R00031